BABY ALGEBRA
FOR BABY AND YOU

ANITA ELAINE HENDERSON-WATSON

To order additional copies of this book, contact:
Xlibris Corporation
1-888-795-4274
www.Xlibris.com
Orders@Xlibris.com

Lesson Plan:

<u>Materials needed</u>: Colorful Art supplies:
glitter, playdoh; colored pencils; pens;
pipe cleaners, construction paper, scissors,
compass, stickers, lamination paper,
resource/research material; journal for
group; graphing paper; jump rope (for
measuring and/ if I jump rope six times in
one second how many in ten? [drt]). cubes
blocks to measure. Google.

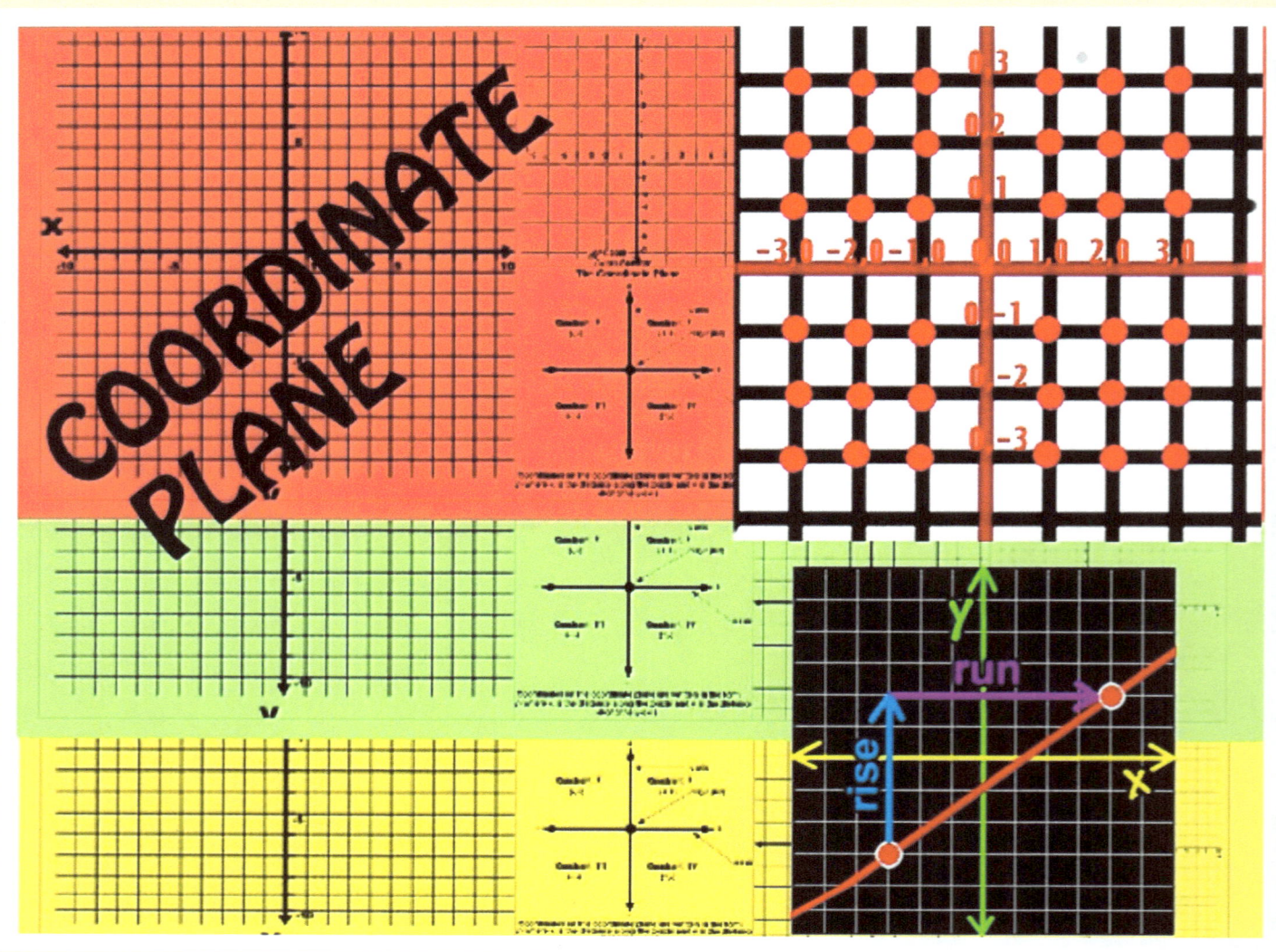

COORDINATE PLANE

Method:

Group work, with each member(6 groups) having responsibility to the whole.

Students make/do with hands on construction of objects that express concepts and notions of the algebra language. Each group prepares a journal. each group is made up of

1. Fact checker;
2. Record Keeper;
3. Planner;
4. Preparer;
5. Consultant;
6. Researcher.

A figure can represent several types simultaneously
Lines Adjacent Vertex Ray Straight line 180 degrees
Y Longitude Lines Meridians Vertical
0-180 Up down N north S south
X Latitude Lines parallels horizontal
N north S south
Left/ Right N/S
0-90 east west

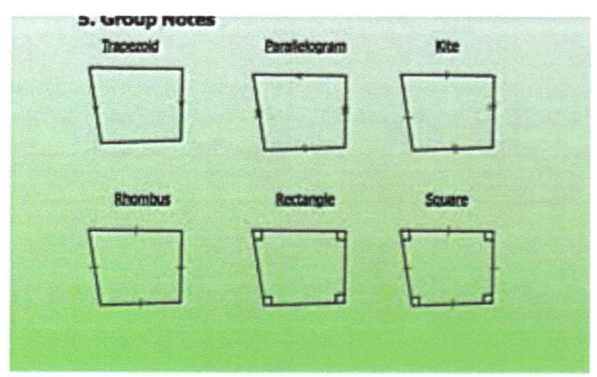

puff puff
little
polygon
prism train that
could   **can**

algebra works

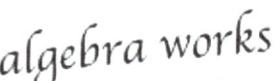

polygon(gon)

pol·y·gon [póllee gòn] many-sided
Flat Surface Like Paper Figure:
a two-dimensional geometric
figure formed of three or more
straight sides Triangle (3-sided)
quadrilateral (4-sided)
pentagon (5-sided),
hexagon (6-sided) and so on.
The word quadrilateral is made of
the words quad (meaning "four")
and lateral (meaning "of sides")

Definitions of cylinder (n)
cyl·in·der [síllindər]   object shaped
                             like tube: an
                              object or
                             shape
                             with straight
                             sides and
                              circular ends
                             of equal size

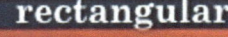

rectangular

C
Y
L
I
N
D
E
R

cylinder

9

SPHERES LINES POLYGON

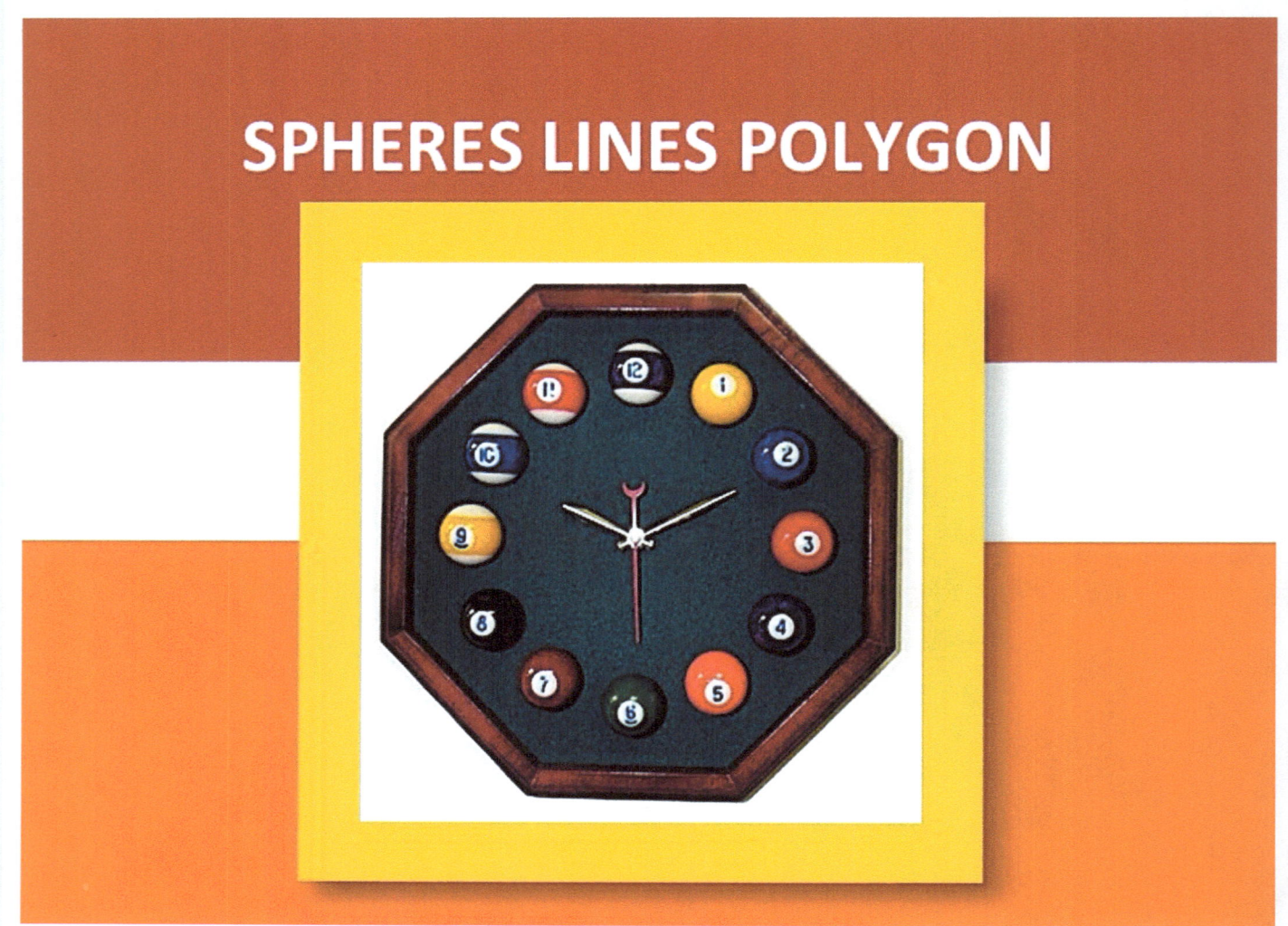

Rectangle
Quadrilateral

cylinder

triangle

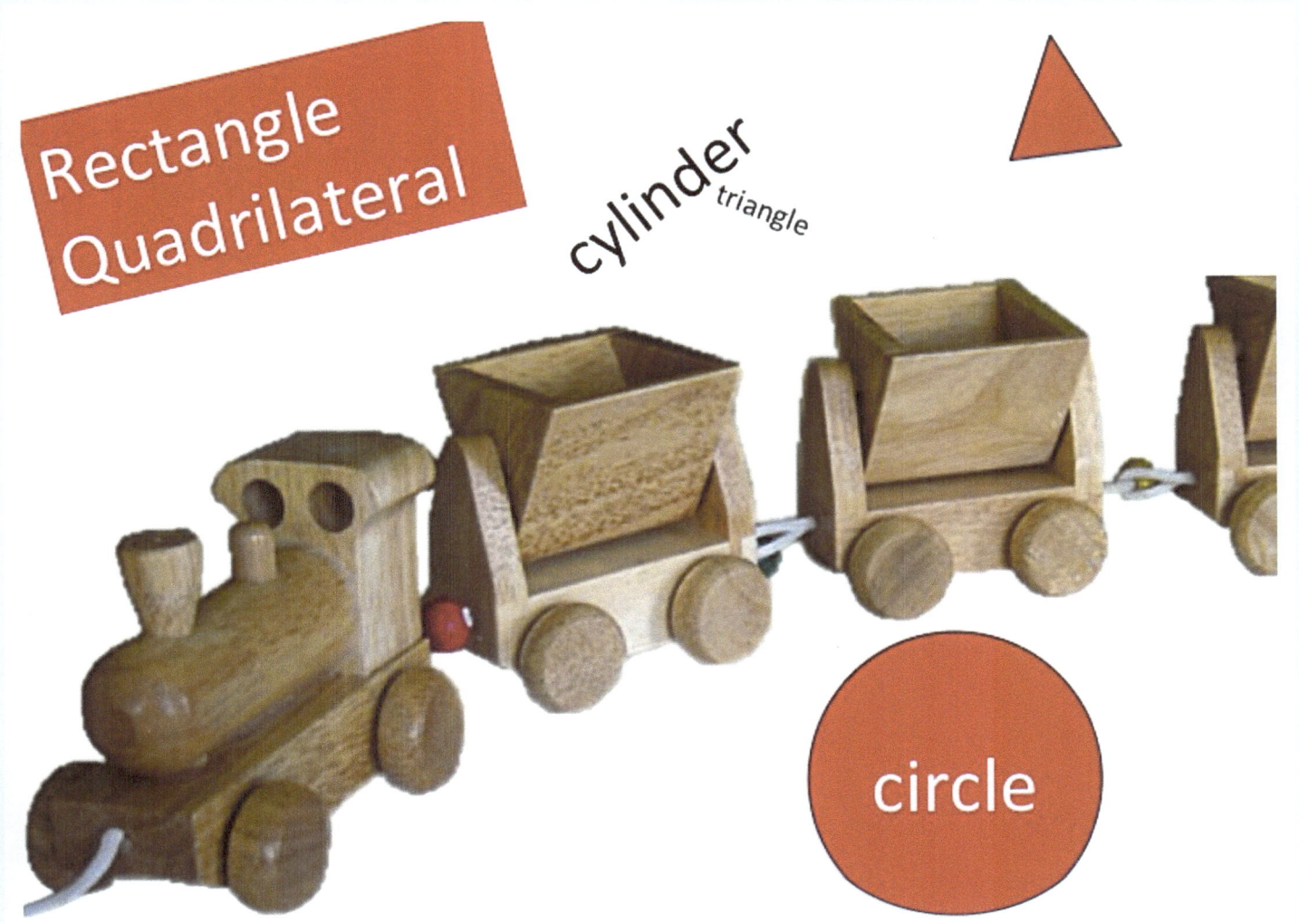

circle

11

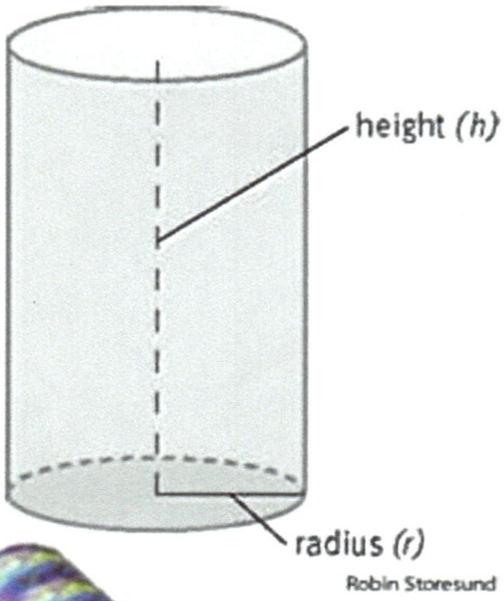

height *(h)*

radius *(r)*

Robin Storesund

Conic Sections

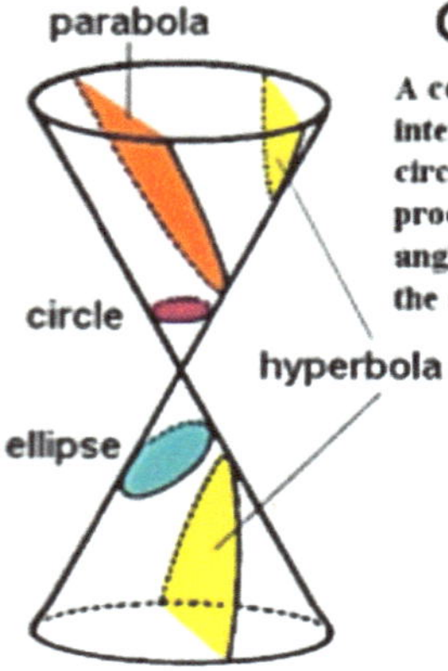

parabola

circle

ellipse

hyperbola

A conic section is formed by the intersection of a plane with a right circular cone. The "kind" of curve produced is determined by the angle at which the plane intersects the surface.

An **ellipse** IS the geometric shape that results from cutting a circular conical or cylindrical surface with an oblique plane (the two unbounded cases being the parabola and the hyperbola).

Campbell's
CONDENSED
TOMATO
SOUP

ORDER 0F OPERATIONS

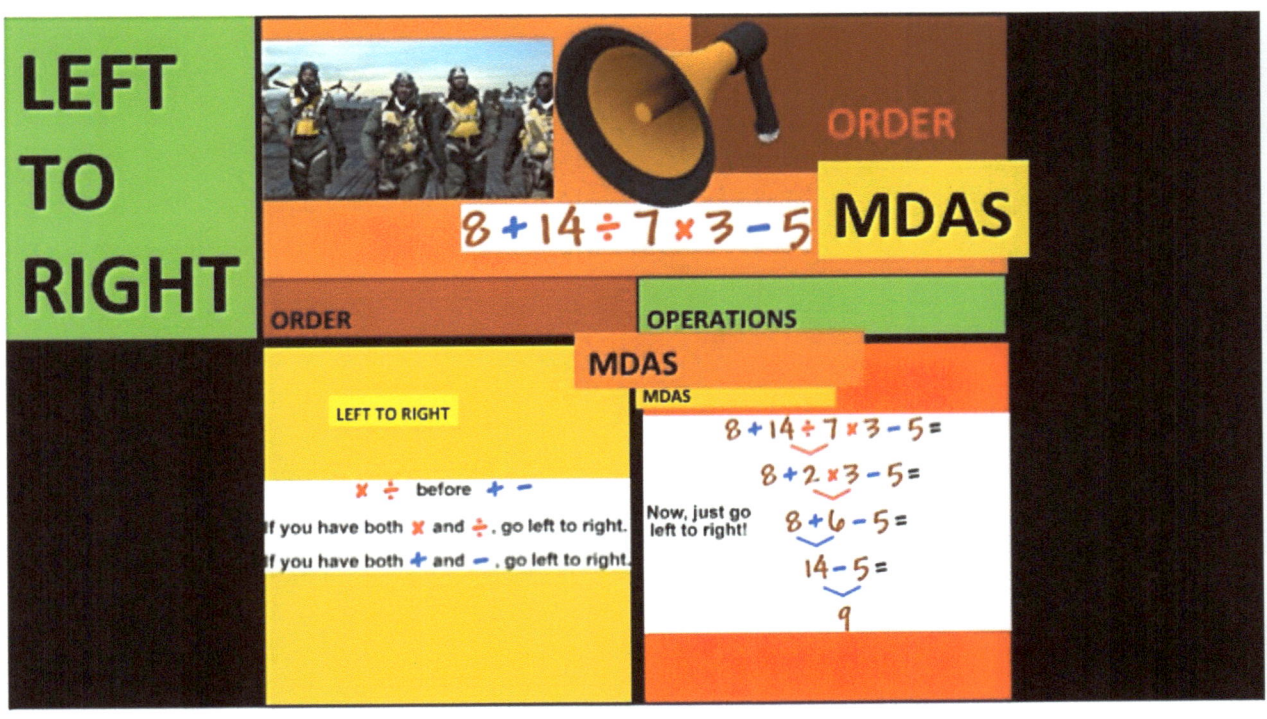

LEFT TO RIGHT

8 + 14 ÷ 7 × 3 − 5

ORDER

MDAS

ORDER

OPERATIONS

MDAS

MDAS

LEFT TO RIGHT

× ÷ before + −

If you have both × and ÷, go left to right.

If you have both + and − , go left to right.

$$8 + 14 \div 7 \times 3 - 5 =$$
$$8 + 2 \times 3 - 5 =$$
$$8 + 6 - 5 =$$
$$14 - 5 =$$
$$9$$

Now, just go left to right!

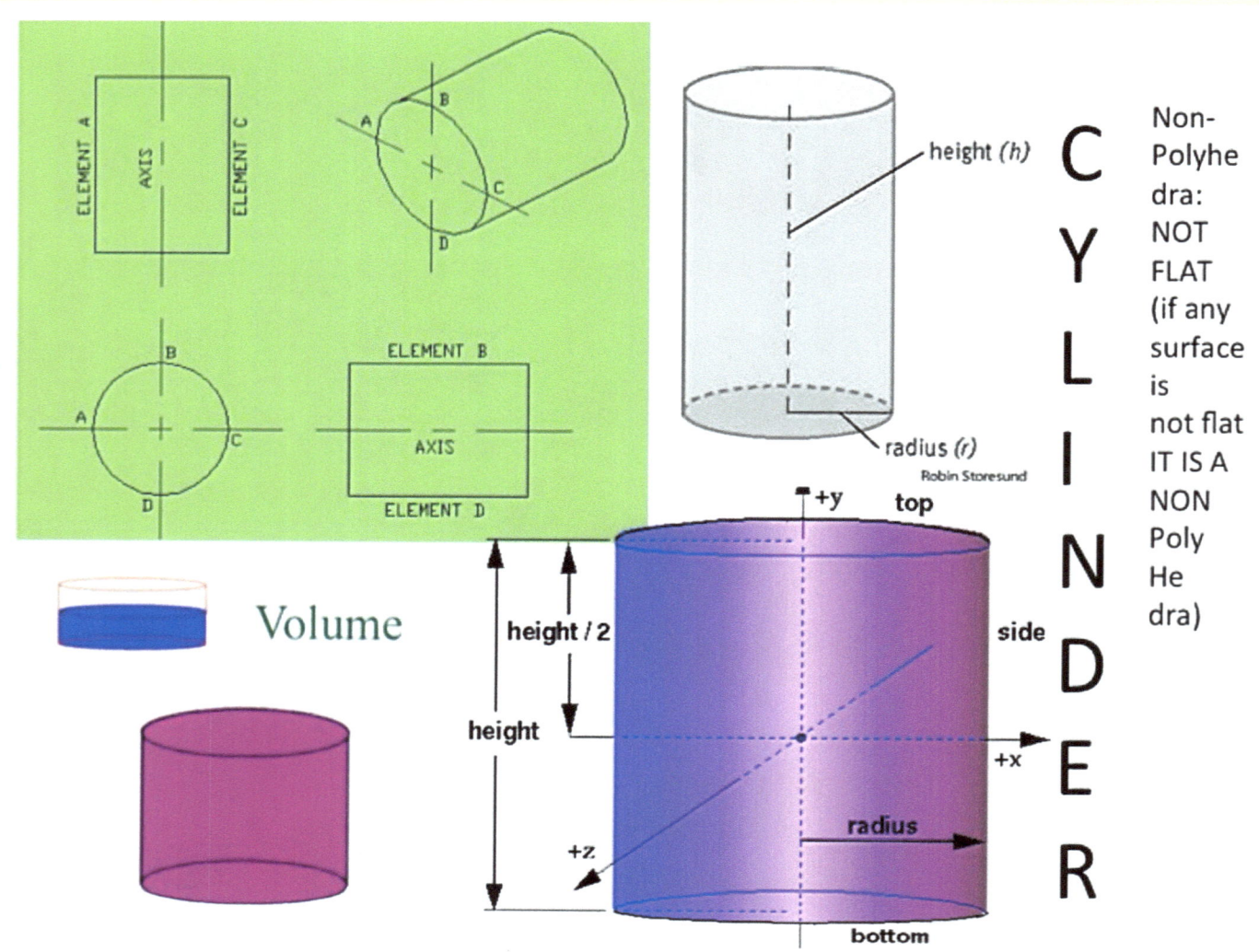

height *(h)*

radius *(r)*

Robin Storesund

Volume

+y   top

height / 2

height

side

+x

+z

radius

bottom

C Y L I N D E R

Non-Polyhe dra: NOT FLAT (if any surface is not flat IT IS A NON Poly He dra)

ELEMENT A

AXIS

ELEMENT C

A

B

C

D

ELEMENT B

AXIS

ELEMENT D

B

A

C

D

14

Classification of Quadrilaterals

Shape	Characteristic	Name
	No sides parallel	Trapezium
	Exactly one pair of parallel sides	Trapezoid
	Two pairs of parallel sides	Parallelogram
	Parallelogram with congruent sides	Rhombus
	Parallelogram with right angles	Rectangle
	Rectangle with congruent sides	Square

Note that squares, rectangles, and rhombuses are types of parallelograms and that a square is a type of rectangle and a type of rhombus.

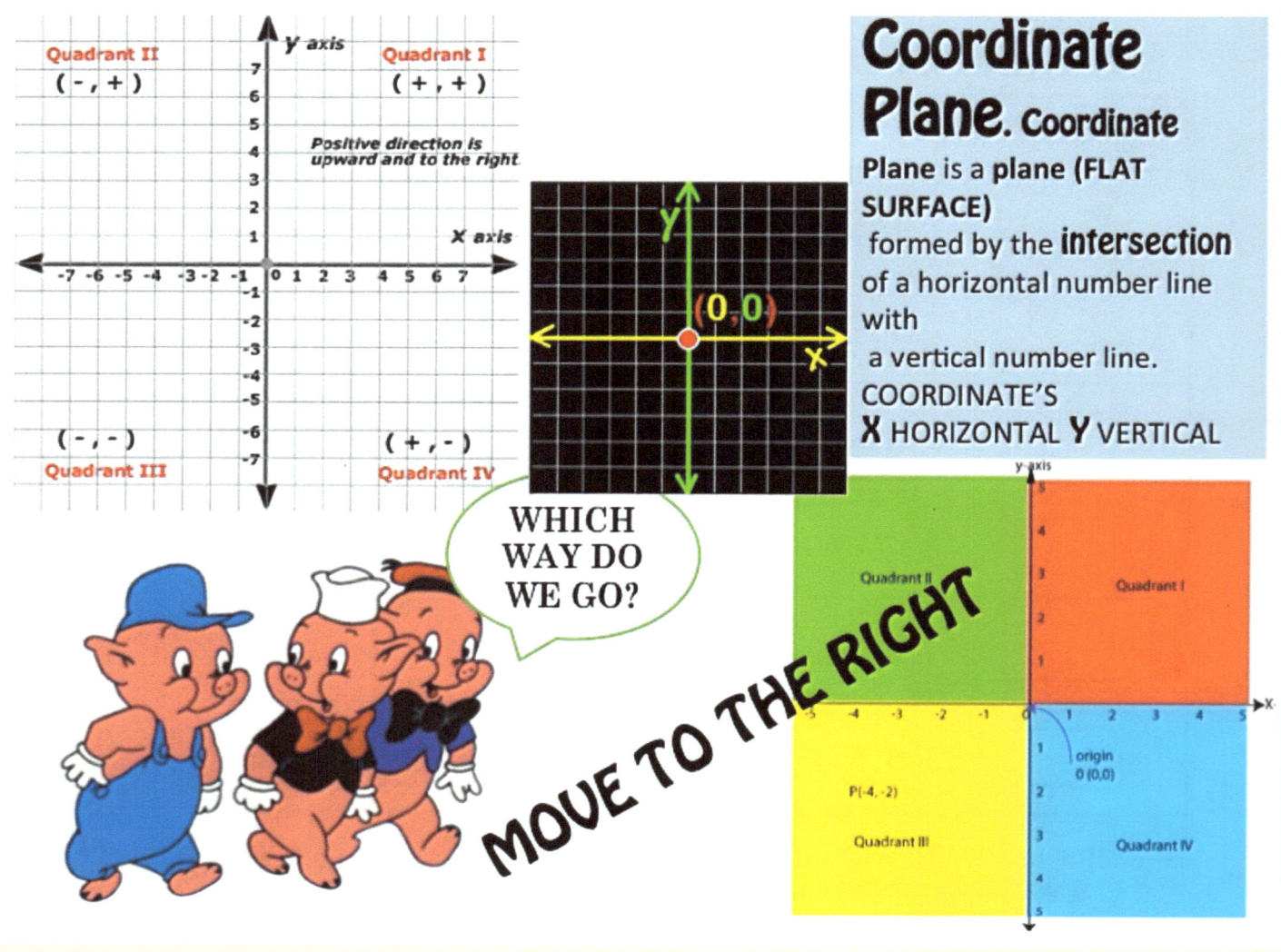

NAME	FIGURE	AREA	PERIMETER CIRCUMFERENCE
TRIANGLE		$A = \dfrac{b \times h}{2}$	$P=MN+NP+PM$
PARALLELOGRAM		$A = b \times h$	$P=DE+EF+FG+GD$
RHOMBUS		$A = b \times h$	$P = b+b+b+b$ $P = 4b$
RECTANGLE		$A = L \times w$	$P = L+w+L+w$ $P = 2L+2w$
SQUARE		$A = l^2$	$P = l+l+l+l$ $P = 4l$
TRAPEZOID		$A = \dfrac{(B+b) \times h}{2}$	$P=MN+NP+PR+RM$
CIRCLE		$A = \pi r^2$	$C = 2\pi r = \pi d$

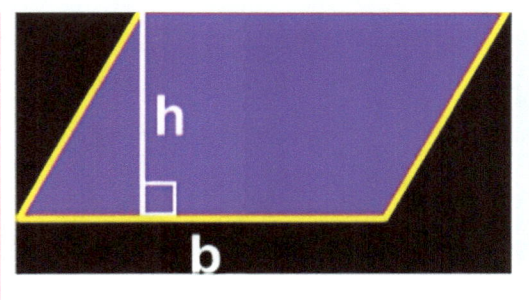

Ordered Pair

An ordered pair is a set of two numbers in the form: (x, y)   Example: (2, -3)

Ordered pairs are used in Cartesian coordinates.

Origin

The origin is the point (0, 0) on the Cartesian plane.

It's called the origin because it's the starting place when you plot a point.

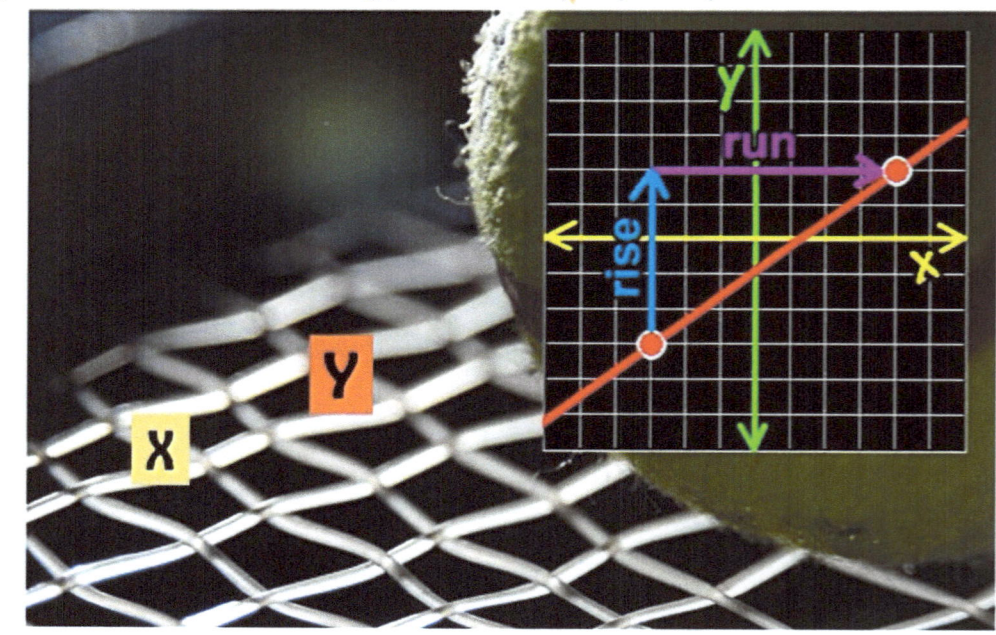

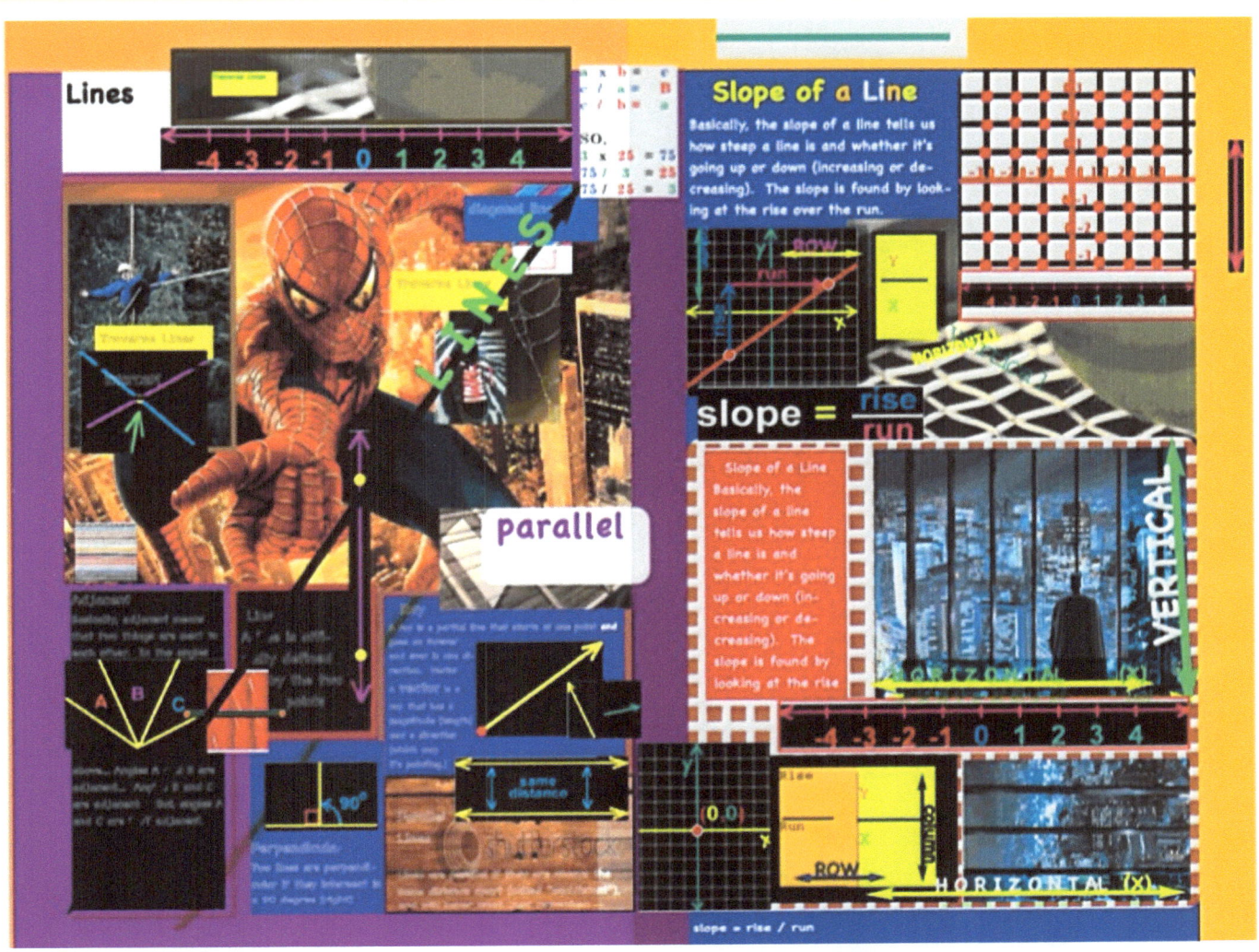

Lines

Slope of a Line

Basically, the slope of a line tells us how steep a line is and whether it's going up or down (increasing or decreasing). The slope is found by looking at the rise over the run.

$$slope = \frac{rise}{run}$$

parallel

Definitions of cone (n)

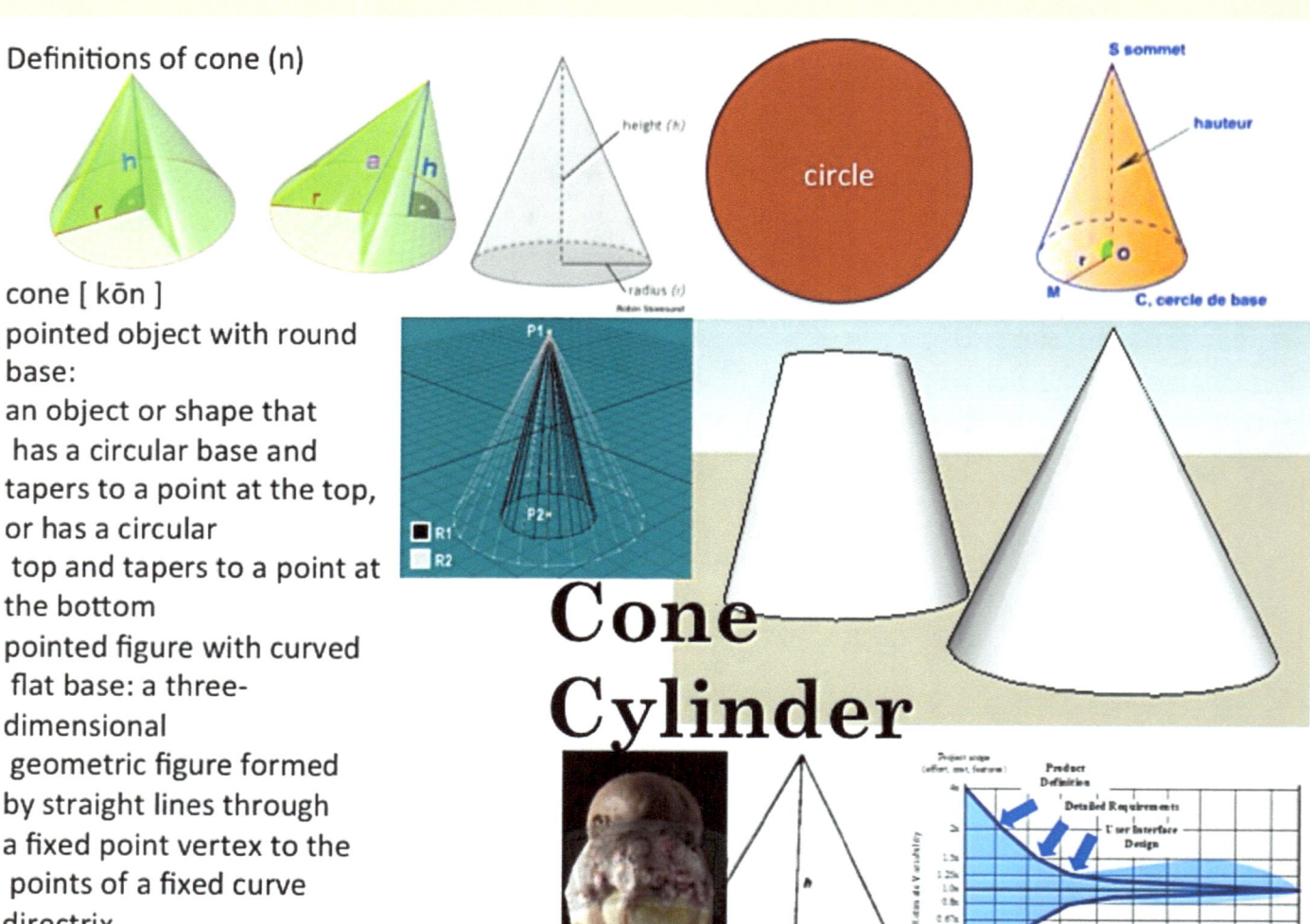

cone [kōn]
pointed object with round
base:
an object or shape that
 has a circular base and
tapers to a point at the top,
or has a circular
 top and tapers to a point at
the bottom
pointed figure with curved
 flat base: a three-
dimensional
 geometric figure formed
by straight lines through
 a fixed point vertex to the
 points of a fixed curve
directrix.

Cone
Cylinder

Definitions of parallel (adj)
par·al·lel [párrə lèl]
always same distance apart:
relating to or being lines,
planes, or curved surfaces that
are always the same distance
apart and therefore never meet

Parallel Lines

par·al·lel·o·gram [pàrrə léllə gràm]
four-sided geometrical figure: a two-dimensional
 geometric figure formed of four sides in which both
 pairs of opposite sides are parallel and of equal length
, and the opposite angles are equal
Synonyms: rhomboid, diamond, lozenge

Sides are outside
Angles are Inside

Par
Al
Lel
ogram

Prism Have Flat Sides

A prism is a polyhedron that is formed with two parallel polygons (the bases - top and bottom) that are connected at the

edges with rectangles. My

example in the picture

Flat Prism Sides

Polyhedra : have flat faces A polyhedron (plural polyhedral or polyhedrons) is often defined as a geometric solid with flat faces and straight edges.

Surface Area

$$A = 2 (wh + lw + lh)$$

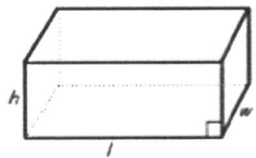

Volume

$$V = lwh$$

Measure me like this!

23

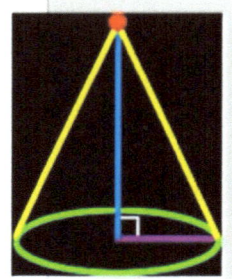

CONE OF LEARNING
WE TEND TO REMEMBER OUR LEVEL OF INVOLVEMENT
(developed and revised by Bruce Hyland from material by Edgar Dale)

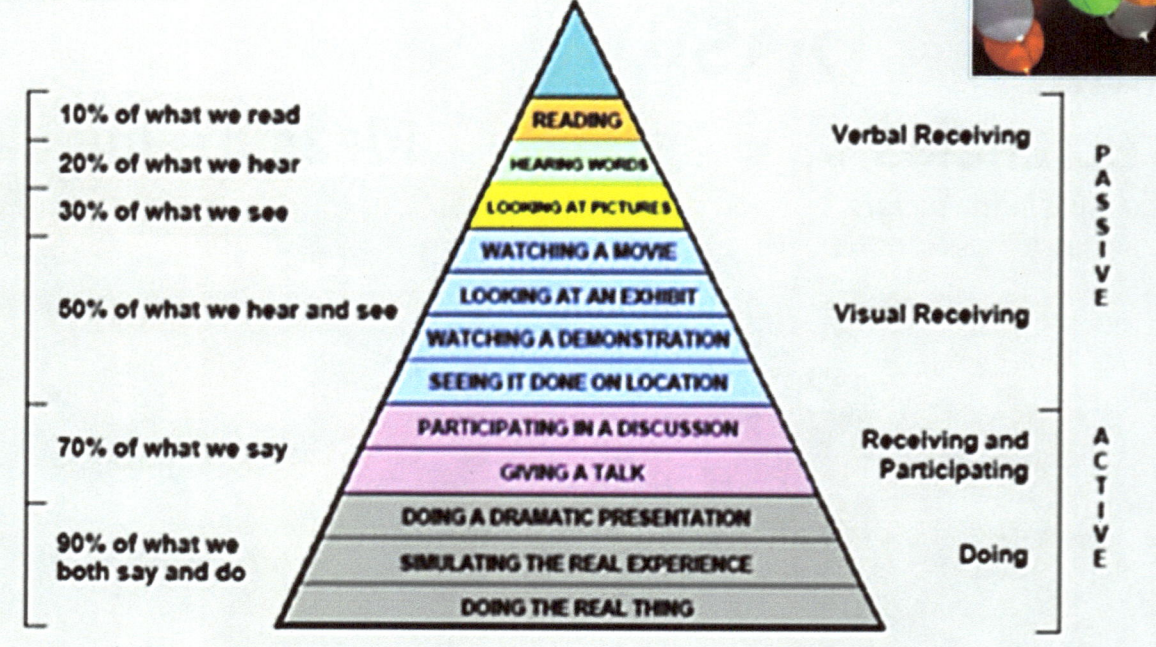

10% of what we read — READING — Verbal Receiving

20% of what we hear — HEARING WORDS

30% of what we see — LOOKING AT PICTURES

WATCHING A MOVIE

50% of what we hear and see — LOOKING AT AN EXHIBIT — Visual Receiving

WATCHING A DEMONSTRATION

SEEING IT DONE ON LOCATION

70% of what we say — PARTICIPATING IN A DISCUSSION — Receiving and Participating

GIVING A TALK

DOING A DRAMATIC PRESENTATION

90% of what we both say and do — SIMULATING THE REAL EXPERIENCE — Doing

DOING THE REAL THING

PASSIVE

ACTIVE

Edgar Dale, Audio-Visual Methods in Teaching (3rd Edition), Holt, Rinehart, and Winston (1969).

Triangles
Triangles

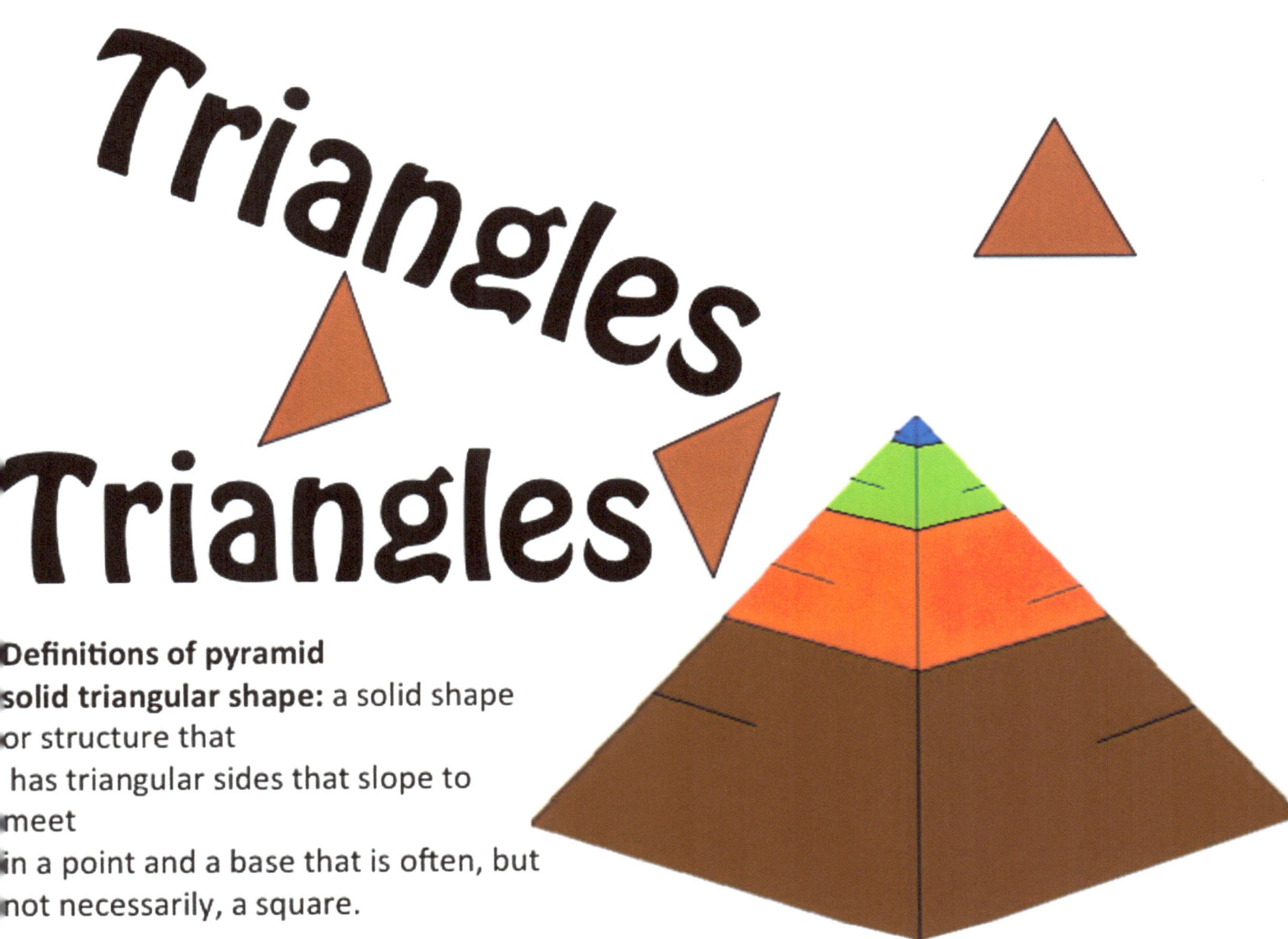

Definitions of pyramid
solid triangular shape: a solid shape or structure that
 has triangular sides that slope to meet
in a point and a base that is often, but not necessarily, a square.

Volume

Sphere

$$V = \frac{\pi d^3}{6}$$

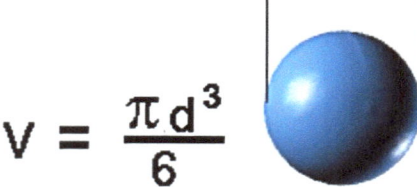

Volume is the three–dimensional space occupied by an object.

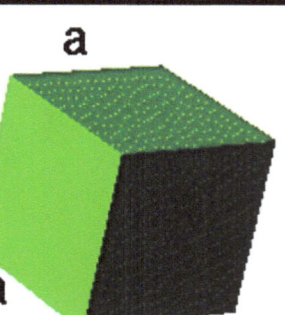

Cube

$$V = a^3$$

$$V = abh$$

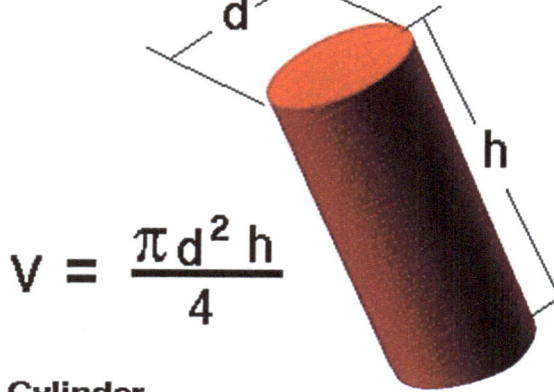

$$V = \frac{\pi d^2 h}{4}$$

Cylinder

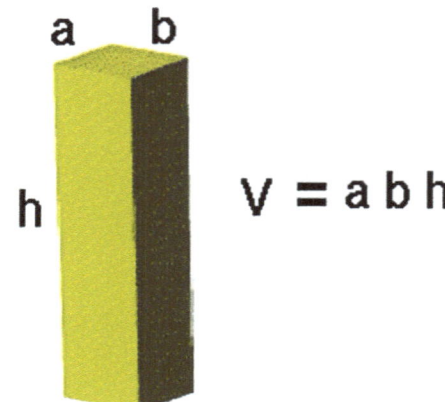

Rectangular Prism

26

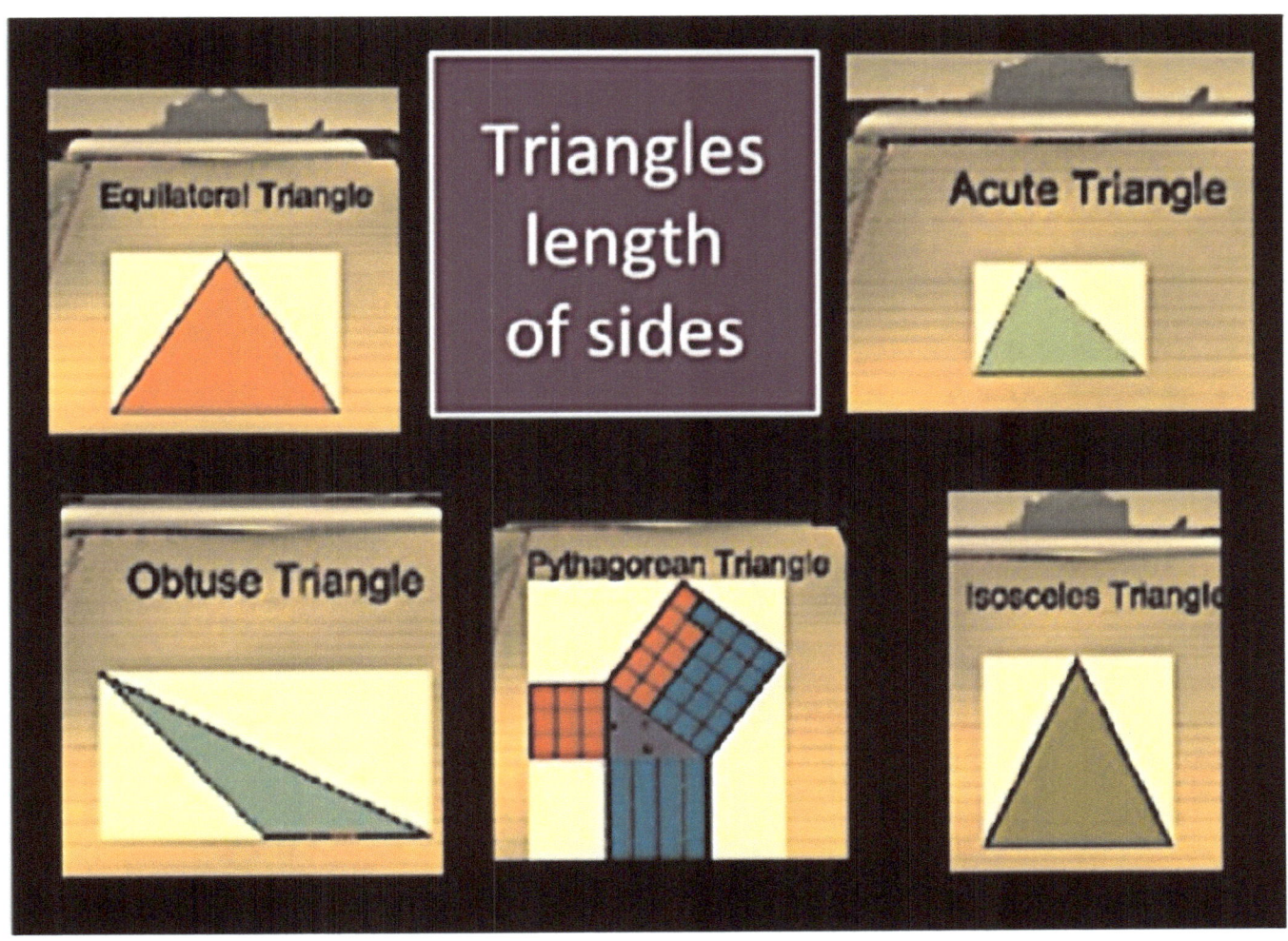

Alternate Interior Angles:
Suppose that L, M, and T are distinct lines. Then L and M are parallel
if and only if alternate interior angles of the intersection of L and T and M and T are equal.

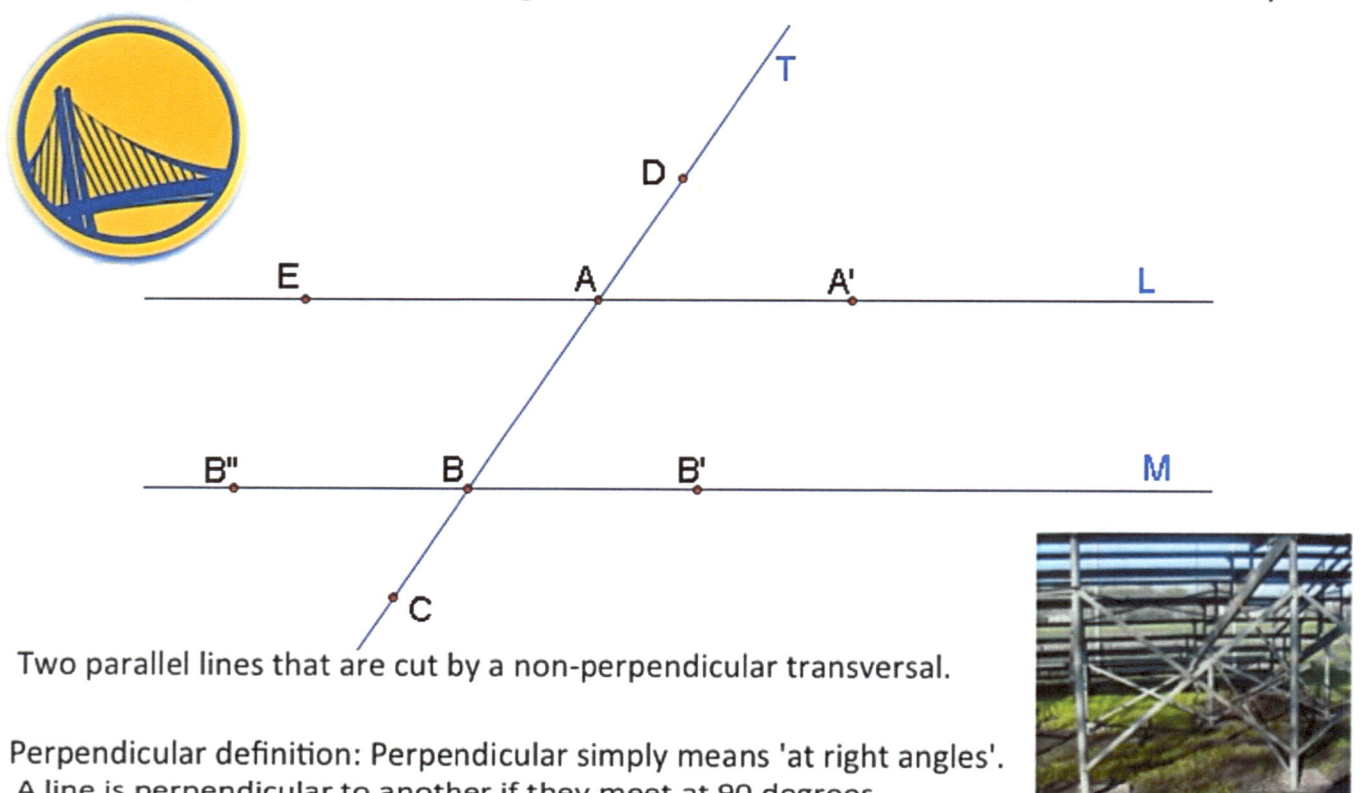

Two parallel lines that are cut by a non-perpendicular transversal.

Perpendicular definition: Perpendicular simply means 'at right angles'.
A line is perpendicular to another if they meet at 90 degrees.

LINES /ANGLES

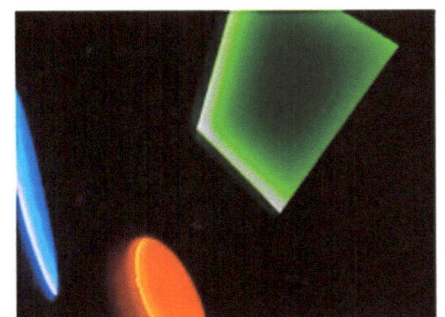

not parallel or
   perpendicular:
Oblique

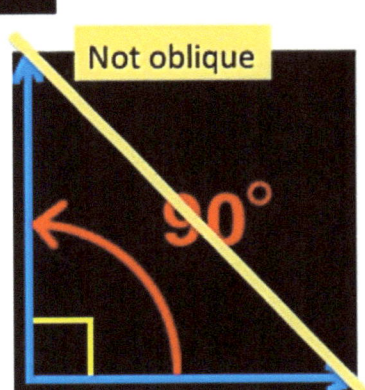

Not oblique

90°

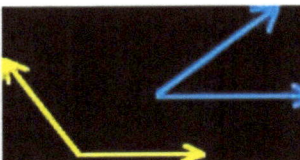

An **angle**, such as an
acute or obtuse
angle,
 that is not a right
angle or a multiple
of a right **angle.**

Oblique is
either plus or
minus 90
degrees

Right Angle
an angle that
is 90° exactly
**Straight
Angle**
an angle that
is 180° exactly

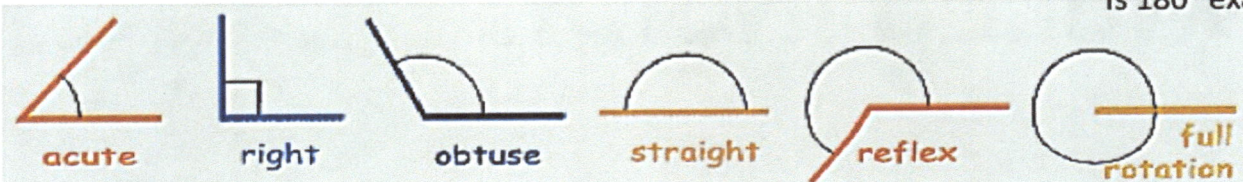

acute   right   obtuse   straight   reflex   full rotation

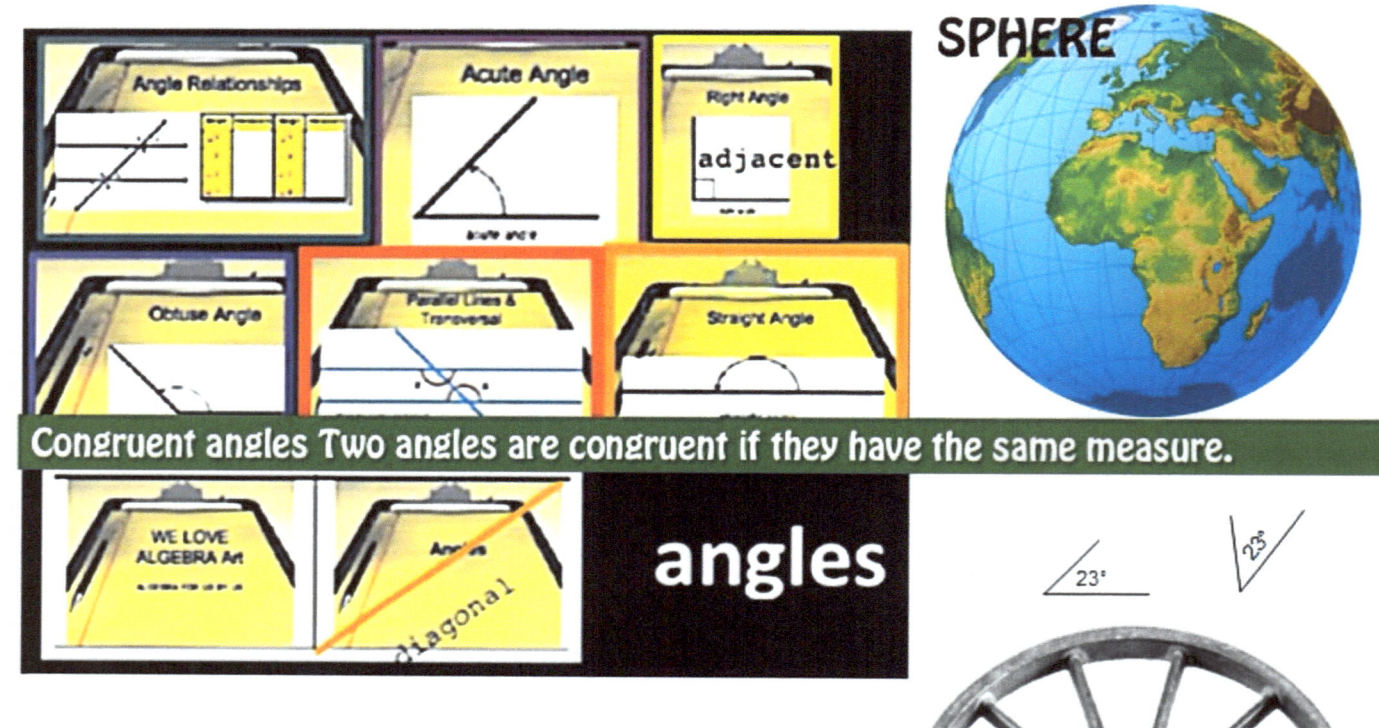

SPHERE

Congruent angles Two angles are congruent if they have the same measure.

angles

23°

23°

ANGLES INSIDE

Reflex Angle an angle that is greater than 180°

Obtuse Angle an angle that is greater than 90° but less than 180

Right Angle
an angle that is 90°exactly

Acute Angle IS an angle that is less than 90°

Straight Angle
an angle that is 180° exactly

Right Angle
an angle that is 90° exactly

and down column Vertical
lattitude north and south
Slope Inclines From
Horizontal "X" right left
east-west row

LONGITUDE

FINDING PATTERNS

Baby Algebra

Lines Linear
Equations Perpendicular
Ray Line Segments
Traverse Bisect Oblique
Slope Angle Slant
Adjacent Parallels Congruent
Similarity Corresponding
Intercept which way? "Y" up
and down column Vertical
lattitude north and south
Slope Inclines From
Horizontal "X" right left
east-west row

LONGITUDE

32

Relativity

a x b = c
c / a = B
c / b = a

SO,
3 x 25 = 75
75 / 3 = 25
75 / 25 = 3

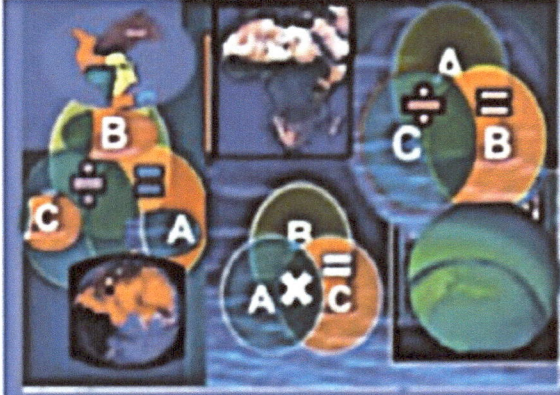

Oblique Angle IS + or – 90 degrees
NOT 90 DEGREES
An oblique angle is an angle that is not a right angle. (An oblique angle is an angle whose measure is not 90 degrees.) + or -

Tennis can be broken down into a game of angles.

Obtuse Angle
An obtuse angle is an angle whose measure is greater than 90 degrees.

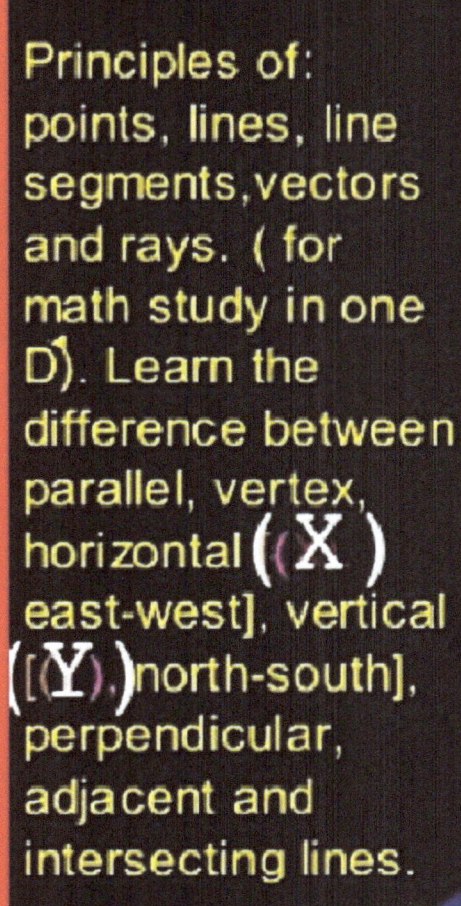

Principles of: points, lines, line segments, vectors and rays. (for math study in one D). Learn the difference between parallel, vertex, horizontal ((X)) east-west], vertical [(Y).)north-south], perpendicular, adjacent and intersecting lines.

lines

Lines
Adjacent
Vertex Ray
Straight
line 180
degrees
Y Longitude
Lines
Meridians
Vertical
0-180 Up
down N
north S
south
X Latitude
Lines
parallels
horizontal N
north S
south Left/
Right N/S
0-90 east

Square
Four
4

Hexagon
six
6

Octagon
eight 8

Dodecagon
Twelve
12

Fill in the blank (Use the above word bank)

1. Percent means how many _____ of 100.

1. Percent means how many out of 100.

2. Percents can be greater than 100%. True or False
2. Percents can be greater than 100%. True or False True

3. To change a percent to a decimal...drop the percent sign and move the decimal two places to the _____.

3. To change a percent to a decimal...drop the percent sign and move the decimal two places to the left.

4. Part/base = rate/100 used to work with proportions and _____
4. Part/base = rate/100 used to work with proportions and percents.

5. In the above statement, _____ – represents the whole amount.
5. In the above statement, base – represents the whole amount.

$$\frac{90}{N} = \frac{20\%}{100\%}$$

$$\frac{90}{N} \bowtie \frac{20\%}{100\%}$$

$$(90)(100) = 20N$$

$$9000 = 20N$$

$$\frac{9000}{20} = N$$

$$450 = N$$

$$\frac{3}{5} \bowtie \frac{6}{10}$$

CROSS MULTIPLY RATIO PROPORTION PERCENTS

$$\frac{4}{1} = \frac{12}{x}$$

Periodic Table of the Elements

1																	18
1 H 1.01	2											13	14	15	16	17	2 He 4.00
3 Li 6.94	4 Be 9.01											5 B 10.81	6 C 12.01	7 N 14.01	8 O 16.00	9 F 19.00	10 Ne 20.18
11 Na 22.99	12 Mg 24.30	3	4	5	6	7	8	9	10	11	12	13 Al 26.98	14 Si 28.09	15 P 30.97	16 S 32.07	17 Cl 35.45	18 Ar 39.95
19 K 30.10	20 Ca 40.08	21 Sc 44.96	22 Ti 47.88	23 V 50.94	24 Cr 52.00	25 Mn 54.94	26 Fe 55.85	27 Co 58.93	28 Ni 58.69	29 Cu 63.55	30 Zn 65.39	31 Ga 69.72	32 Ge 72.61	33 As 74.92	34 Se 78.96	35 Br 79.90	36 Kr 83.80
37 Rb 85.47	38 Sr 87.62	39 Y 88.91	40 Zr 91.22	41 Nb 92.91	42 Mo 95.94	43 Tc (97.91)	44 Ru 101.07	45 Rh 102.91	46 Pd 106.42	47 Ag 107.87	48 Cd 112.41	49 In 114.82	50 Sn 118.71	51 Sb 121.75	52 Te 127.60	53 I 126.90	54 Xe 131.29
55 Cs 132.91	56 Ba 137.33	57 La 138.91	72 Hf 178.49	73 Ta 180.95	74 W 183.85	75 Re 186.21	76 Os 190.23	77 Ir 192.22	78 Pt 195.08	79 Au 196.97	80 Hg 200.59	81 Tl 204.38	82 Pb 207.2	83 Bi 208.98	84 Po (208.98)	85 At (209.99)	86 Rn (222.02)
87 Fr (223.02)	88 Ra (226.03)	89 Ac (227.03)	104 Rf (261.11)	105 Ha (262.11)	106 Sg (263.12)												

58 Ce 140.12	59 Pr 140.91	60 Nd 144.24	61 Pm (144.91)	62 Sm 150.36	63 Eu 151.97	64 Gd 157.25	65 Tb 158.93	66 Dy 162.50	67 Ho 164.93	68 Er 167.26	69 Tm 168.93	70 Yb 173.04	71 Lu 174.97
90 Th 232.04	91 Pa 231.04	92 U 238.03	93 Np (237.05)	94 Pu (244.06)	95 Am (243.06)	96 Cm (247.07)	97 Bk (247.07)	98 Cf (251.08)	99 Es (252.08)	100 Fm (257.10)	101 Md (258.10)	102 No (259.10)	103 Lr (262.11)

Please Excuse My Dear Aunt

Order of Operations

Order of operations tells us what order we are supposed to do things in a math problem. For example,

what's the answer to this? 4 + 10 x 2

Do we do the 4 + 10 first? Or the 10 x 2 first?

Here is the official order of operations:

Remember (1) Parenthesis
(2) Exponents
(3) Multiplication & Division
(4) Addition & Subtraction

So, for 4 + 10 x 2, we do multiplication, then addition... 4 + 10 x 2 = 4 + 20 = 24.

A common phrase to remember the order is:
Please Excuse My Dear Aunt

8. 3:4 is the same as 3 **to** 4

8. The above ratio can be expressed as 3:4, 3 to 4, or ¾.

9. The denominator for percents when represented as a fraction is always _____.

9. The denominator for percents when represented as a fraction is always 100. In order to help students remember this, I ask the following question. What percentage of the problems do you want to get correct on the next test? Of course, the answer is 100%. However, if you grade was an 83. It means used answered 83 out of 100 or 83/100 or 83% correct.

11. _____ - comparison of two amounts

11. Ratio - comparison of two amounts

12. _____ rate

12. Unit rate

Unit rate – the amount compared to one; in a ratio, the second term would be one. For instance, a box of 12 pens cost $24.00. One pen or one unit cost $2. How did I get $2? Divide 24 by 12.

Percent of change

(original amount – new amount) / original amount

10. Percent of _____ = (original amount – new amount) / original amount

10. Percent of change = (original amount – new amount) / original amount

Percent of change means the amount a rate has changed over a period of time.

DR. ANGELA DAVIS

Civil and Human Rights in a Democracy

April 9, 7:00 PM | Jesse Auditorium

MU Students: Free | Public: $5

ACTIVIST. EDUCATOR. PHILOSOPHER.

/ =   / =

DIVIDE DIVIDE

BY   BY

KEY WORDS

Lessons HAVE NO Grade Level

Early Childhood Lessons

Elementary Lessons

Jr. High/Middle School

High School Lessons

Undergraduate Lessons

Elementary Substitute

Middle School - Substitute

Lesson Idea Pages

Drama and Art

Perimeter formula
Square   4 * side
Rectangle   2 * (length + width)
Parallelogram   2 * (side1 + side2)
Triangle side1 + side2 + side3
Regular n-polygon   n * side
Trapezoid   height * (base1 + base2) / 2
Trapezoid   base1 + base2 + height * [csc(theta1) + csc(theta2)]
Circle   2 * pi *   radius
Ellipse   4 * radius1   * E(k,pi/2)
E(k,pi/2) is the Complete   Elliptic Integral of the Second Kind
k = (1/radius1) *   sqrt(radius1^2 - radius2^2)

Area formula
Square   side^2
Rectangle   length * width
Parallelogram   base * height
Triangle base * height / 2
Regular n-polygon   (1/4) * n * side^2 * cot(pi/n)
Trapezoid   height * (base1 + base2) / 2
Circle   pi * radius^2
Ellipse   pi * radius1 * radius2
Cube (surface)   6 * side^2

Sphere (surface) $4 * pi * radius^2$
Cylinder (surface of side) perimeter of circle * height
    $2 * pi * radius * height$
Cylinder (whole surface)   Areas of top and bottom circles + Area of the side
    $2(pi * radius^2) + 2 * pi * radius * height$
Cone (surface) $pi * radius * side$
Torus (surface) $pi^2 * (radius2^2 - radius1^2)$

Volume formula

Cube $side^3$
Rectangular Prism $side1 * side2 * side3$
Sphere $(4/3) * pi * radius^3$
Ellipsoid $(4/3) * pi * radius1 * radius2 * radius3$
Cylinder $pi * radius^2 * height$
Cone $(1/3) * pi * radius^2 * height$
Pyramid $(1/3) * (base\ area) * height$
Torus $(1/4) * pi^2 * (r1 + r2) * (r1 - r2)^2$

Regular n-polygon $(1/4) * n * side^2 * cot(pi/n)$
Trapezoid $height * (base1 + base2) / 2$
Circle $pi * radius^2$
Ellipse $pi * radius1 * radius2$
Cube (surface) $6 * side^2$

```
F N T F K O L A M T R U E
P I Z I Q N X L U K P K S
R W F Q G Y S O U N I T W
O A O T G R L L P W N N H
P U R V Y Q S W A E E B O
O D D D G A T P C F J H L
R V E G C V T R C X C U E
T F R J M H E N Z U C N P
I B J X W P A K J P A D V
O R A T I O F N V Y T R H
N G F S Z P T S G F O E V
S Y Q M E V B O E E M D P
C R O S S M U L T I P L Y
```

Beverly Mackie'S CROSS MULTIPLY RATIO PROPORTION PERCENTS

PERCENTS -
RATIOS -
PROPORTIONS –
WORD SEARCH
BASE
CHANGE
CROSS MULTIPLY
FIFTY
HUNDRED
LEFT
ORDER
OUT
PERCENTS
PROPORTIONS
RATIO
TO
TRUE
UNIT

POLYNOMIALS

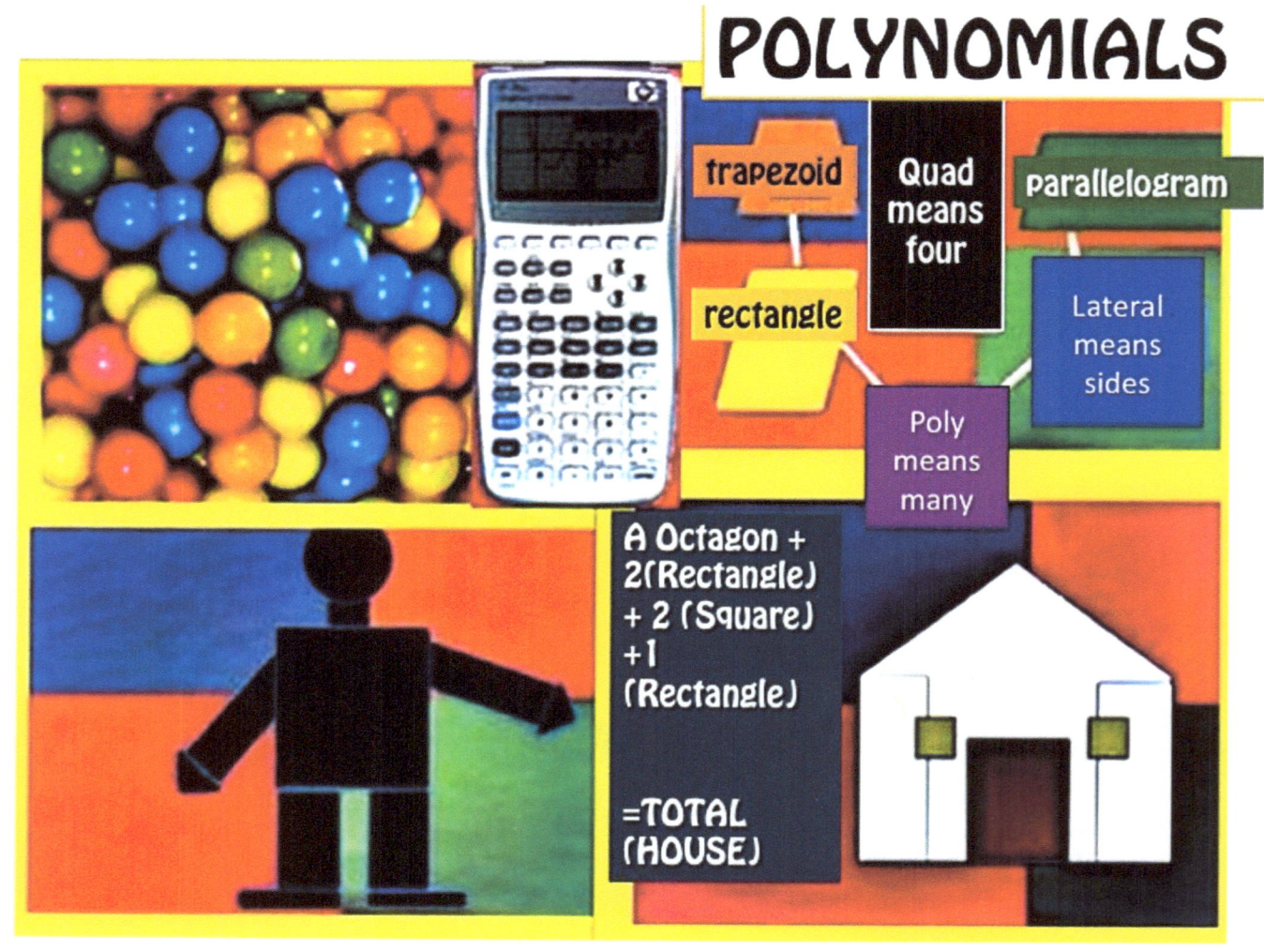

trapezoid

Quad means four

parallelogram

rectangle

Lateral means sides

Poly means many

A Octagon +
2(Rectangle)
+ 2 (Square)
+1
(Rectangle)

=TOTAL
(HOUSE)

ellipse

6. Two for one is the same as _____ percent.

6. Two for one is the same as fifty percent.

If a calculator costs ten dollars, two calculators would could twenty dollars.
However, "two for one" means
I can buy two calculators for the price of one calculator. In other words,
I pay a total of ten dollars for both calculators. That's the same as paying
half price or five dollars for each one. What a deal! Half price is equivalent to fifty
percent.

7. Part / _____ represents a fraction

7. Part / whole represents a fraction.

By definition, a fraction represents a part of a whole. Some teachers may say a fraction represents
part of a total. Let's say you ordered a taco pizza. When the pizza arrives it has
been cut into eight slices with no slices eaten. At this point, it is a whole pizza. Since
your brother fell asleep while waiting on the pizza, you decide to get a head start before
waking him up. So, you eat three slices of pizza! Your brother catches you wiping your
mouth after finishing the third slice. He accuses you of eating three parts of the whole
pizza, or 3/8 of the pizza! You are guilty as charged.

Social Studies
4. State (verbal and/or written) what you know about algebra describing, if any, your feelings. What are your thoughts, if any? Recall any experience that you have had, any story, make believe, or real. Please give examples, grasp them, restate and conclude.

Language Arts
5. Group similar algebraic terms in thesaurus format, for example synonymous algebraic terms - a figure can represent more than one type/classification.

LINES /ANGLES

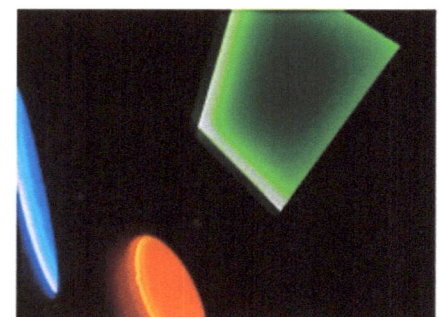

not parallel or
 perpendicular:
Oblique

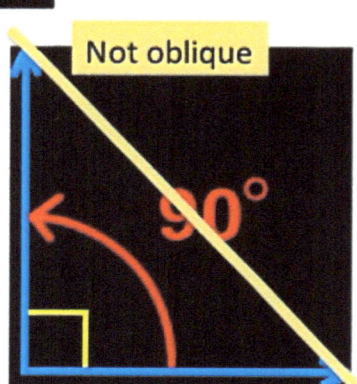

Not oblique

90°

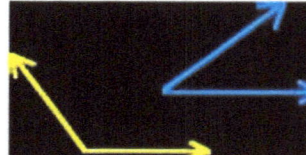

Oblique is
either plus or
minus 90
degrees

An **angle**, such as an
acute or obtuse
angle,
 that is not a right
angle or a multiple
of a right **angle**.

Right Angle
an angle that
is 90° exactly
**Straight
Angle**
an angle that
is 180° exactly

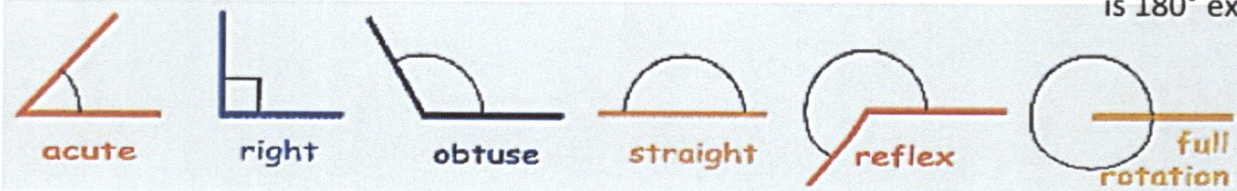

acute   right   obtuse   straight   reflex   full rotation

Midpoint   The midpoint of the segment (x1, y1) to (x2, y2)

The midpoint (also known as class mark in relation to histogram)
is the middle point of a line segment. It is equidistant from both endpoints.

Formulas

The formula for determining the midpoint of a segment in the plane,
with endpoints (x1) and (x2) is:

$$\frac{x_1 + x_2}{2}$$

The formula for determining the midpoint of a segment in the plane,
with endpoints (x1, y1) and (x2, y2) is:

$$\left(\frac{x_1 + x_2}{2}, \frac{y_1 + y_2}{2} \right)$$

The formula for determining the midpoint of a segment in the space, with
endpoints (x1, y1, z1) and (x2, y2 z2) is:

$$\left(\frac{x_1 + x_2}{2}, \frac{y_1 + y_2}{2}, \frac{z_1 + z_2}{2} \right)$$

More generally, for an n-dimensional space with axes , the midpoint of an
interval is given by:

Construction

$$\left(\frac{x_{1_1} + x_{1_2}}{2}, \frac{x_{2_1} + x_{2_2}}{2}, \frac{x_{3_1} + x_{3_2}}{2}, \ldots, \frac{x_{n_1} + x_{n_2}}{2} \right)$$

The midpoint of a line segment can be located by first constructing a lens
using circular arcs, then connecting the cusps of the lens.
The point where the cusp-connecting line intersects the segment is then the midpoint.
It is more challenging to locate the midpoint using only a compass, but it is still possible.

Golden rectangle
• Parallelogram
• Rhombus
• Trapezoid
• Trapezoid median
• Trapezium
• Kite
Area of various polygon types
• Regular polygon area
• Irregular polygon area
• Rhombus area
• Kite area
• Rectangle area
• Area of a square
• Trapezoid area
• Parallelogram area
Perimeter of various poly-

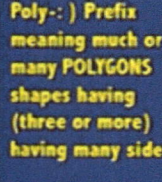

Poly-:) Prefix meaning much or many POLYGONS shapes having (three or more) having many sides

gon types
• Perimeter of a polygon (regular and irregular)
• Perimeter of a triangle
• Perimeter of a rectangle
• Perimeter of a square
• Perimeter of a parallelogram
• Perimeter of a rhombus
• Perimeter of a trapezoid
• Perimeter of a kite
Angles associated with polygons

• Exterior angles of a polygon
• Interior angles of a polygon
• Relationship of interior/exterior
• Polygon central
Tri-angle 3 sides
• Tetragon, 4 sides
• Pentagon, 5 sides
• Hexagon, 6 sides
• Heptagon, 7sides
• Octagon, 8 sides
Nonagon
Enneagon, 9 sides
• Decagon, 10 sides

Goal: Learn to work in cooperative groups, to speak, understand and use 100 words of the algebra language and connecting math to other school subjects.

Objectives:

Language Arts
1. Spelling Test for all algebra specific terms;
2. Vocabulary Definitions own words/ pictures

Arts and Crafts
3. Create an art project that demonstrates concepts depicted by the particular language of algebra.

Symbol

The symbol for congruence is

≅

LINES/ ANGLES

≅

Two angles are **congruent** if they have the same measure.

In geometry, **adjacent** angles, often shortened as adj. ∠s, are angles that have a common ray coming out of the vertex going between two other rays.(*line with direction to infinit*). In other words ... In geometry, two **lines** or planes (or a **line** and a plane) are considered **perpendicular** (or orthogonal) to each other if they form congruent adjacent angles (a **T-shape**). In geometry, two **figures** are **congruent** if they have the same shape and size. Two **line segments** are congruent if they have the same length. But they need not lie at the same angle or position on the **plane**

TABLE OF CONTENTS

Triangle
three 3

Square
Four
4

Pentagon
Five
5

Hexagon
six
6

Heptagon
seven 7

Octagon
eight 8

Decagon
ten
10

Dodecagon
Twelve
12

CONGRUENT = SAME

Two circles are congruent if they have the same size. The size can be measured as the radius, diameter or circumference. They can overlap.

Congruent polygons have an equal number of sides, and all the corresponding sides and angles are congruent. However, they can be in a different location, rotated or flipped over.

Two angles are congruent if they have the same measure.

23°   23°

Venn Diagram

parallelograms

rectangles

squares

rhombuses

Quadrilaterals

trapezoids

NOTE: Similar polygons which be in the same proportions but different sizes.

56